IMAGES
of Aviation

NAVAL AIR STATION
PENSACOLA

This Curtiss flying boat ushers in a new era for the United States Navy as it flies over Pensacola around 1915. Note the masts of the ships in the background. (Courtesy of National Naval Aviation Museum.)

ON THE COVER: The officers and men of Naval Air Station Pensacola pose with the F5L flying boat, a twin-engine plane built at the Naval Aircraft Factory in Philadelphia in the 1920s. (Courtesy of National Naval Aviation Museum.)

IMAGES
of Aviation

NAVAL AIR STATION
PENSACOLA

Maureen Smith Keillor and
AMEC (AW/SW) Richard P. Keillor, MTS

ARCADIA
PUBLISHING

Published by Arcadia Publishing
Charleston, South Carolina

Printed in the United States of America

Library of Congress Control Number: 2013938500

For all general information, please contact Arcadia Publishing:
Telephone 843-853-2070
Fax 843-853-0044
E-mail sales@arcadiapublishing.com
For customer service and orders:
Toll-Free 1-888-313-2665

Visit us on the Internet at www.arcadiapublishing.com

*This book is dedicated to the amazing men and women
who were willing to be launched into the sky in fragile
boxes of wood and fabric and, later, in tin cans; and to the
men and women who keep those craft flightworthy.*

CONTENTS

ACKNOWLEDGMENTS

First, to Evelyn Wheeler and Sue VerHoef, thanks for getting the ball rolling and being great cheerleaders. You are the best research assistants anywhere!

Thank you, to Dr. Rebecca Godlasky, for cementing the foundation.

The marvelous staff at the National Naval Aviation Museum (NNAM) Foundation has been very helpful and encouraging. We could not have done this without their assistance, especially the wonderful gentlemen who staff the Emil Buehler Library. Unless otherwise noted, all photographs in this book come from the NNAM.

Jacquelyn Wilson, archivist at the Pensacola Historical Society, was very helpful, along with her assistant, Carolyn Prime. Shari Johnson, electronic resources/business librarian, and Dean DeBolt, university librarian, Special Collections and West Florida Archives, John C. Pace Library, University of West Florida, were both invaluable in providing access to the *Pensacola News Journal* on microfilm. Peggy Vignolo, administrative office, West Florida Public Library Genealogy Branch, and her volunteer, Bruce Rova (retired Navy), provided excellent resources and suggestions. Cathy Whitney, director, community relations, Public Affairs Office, NAS Pensacola, provided up-to-date statistics and information.

Lt. Katie Kelly, Public Affairs Office, Blue Angels, was most helpful, providing photographs and caption assistance on the Blue Angels chapter.

Finally, and most significantly, we dedicate this effort to our family, for being our biggest cheerleaders, and for their patience and support.

INTRODUCTION

Although there is evidence of human activity in the Pensacola area for over 10,000 years, there has been a European military presence nearly continually since 1559, when Don Tristan de Luna attempted to establish a Spanish base there. For the Spanish, the attraction of the Gulf coast was threefold: a port to help in the defense of the treasure fleets; a means of keeping the French out of the area; and a base from which to spread the gospel to the inhabitants, as there were Dominican priests in de Luna's company. Unfortunately for them, a month after landing, a severe hurricane killed many of the settlers and destroyed most of their ships, including supplies. After nearly two years of severe deprivations, the remainder of the company decided that the area was not even fit for their nearest competitors, the French, and abandoned the area they had named "Polanco." It was not until 1686 that the bay was "rediscovered" by Juan Jordan Reina, who had been leading an expedition searching for the mouth of the Mississippi River. According to William S. Coker in the book *The New History of Florida*, Reina described the Pensacola bay as "the best I have ever seen in my life." Again, European politics and the struggle for dominance in the New World dictated events. The Spanish crown ordered fortifications built, and in 1698, Fort San Carlos was begun on what is today part of Naval Air Station Pensacola. From its time as a Spanish possession until 1821, when both East and West Florida were handed over to the United States as a result of the Adams-Onis Treaty, the flags of three nations flew over Pensacola—Spanish, French, and British—in a military and diplomatic tug-of-war.

As soon as the Stars and Stripes flew over the Florida Territory, Andrew Jackson, as its first appointed military governor, proposed Pensacola as a site for a naval installation. On the heels of the War of 1812, which ended in 1815, Jackson recognized the vulnerability of Florida's coastline. Additionally, poor roads and the lack of railroads isolated Florida from the rest of the nation. Congress approved the construction of a naval base in Pensacola in 1824, and in 1825, Secretary of the Navy Samuel Southard sent a panel of Navy captains—William Bainbridge, Lewis Warrington, and James Biddle—to choose the site. Additionally, in 1828, Pres. John Quincy Adams authorized the purchase of land east of Pensacola along the "Old Federal Road," which had been constructed between Pensacola and Florida's other city, St. Augustine. Today, the area, maintained by the National Park Service, proclaims itself as "America's first tree farm." On the land grew the beautiful live oaks that not only gave the Naval Live Oaks area its name, but provided the dense wood that gave ships like the USS *Constitution* their "ironsides" quality. Little was accomplished at the Navy yard for many years, until the threat of war with Mexico. Then, it was provided with supplies and equipment for the repair and outfitting of the fleet. However, during this time, the US Department of Engineers had been at work establishing fortifications to Pensacola's harbor, building Fort Redoubt to the west of the area selected for the yard, Fort Pickens to the south on Santa Rosa Island, and Fort McRee on Perdido Key, west of Santa Rosa. In addition, the old Spanish fort, San Carlos de Barrancas, was enlarged.

After the war with Mexico, this situation reversed itself. The forts stood empty, and the Pensacola Navy Yard enjoyed a period of intense activity. The focus was on shipbuilding, using the dense wood of the plentiful live oaks in the area. War again rearranged affairs at the yard. Because of its strategic location in the South, Pensacola Navy Yard nearly took on the role of Fort Sumter in the Civil War. US Army lieutenant Adam J. Slemmer wisely opted to consolidate the forces of the four forts within the walls of Fort Pickens. A combination of politics, family loyalties, and bad weather staved off the Confederate attack until after Fort Sumter had already become the flashpoint of the war. The story is told of William Conway, an old quartermaster of 40 years' service in the Navy: Confederate forces entered the Navy yard to occupy it, ordering Conway to lower the flag. According to the account given at the dedication ceremony of a plaque in his honor, Conway said, "I will not do it, sir! That is the flag of my country under which I have served many years. I love it; and will not dishonor it by hauling it down now." Conway was promptly thrown in irons, but he was later celebrated as a hero in his native Maine. The yard was held by the Confederates for several months, during which time more than one battle took place across the bay between Fort Pickens and the Navy yard. But the yard was abandoned when it was realized that the cost of holding it was higher than its value, since it was of no use as a base of naval operations. Getting out of the harbor would be impossible for the Confederates under the guns of Union-held Fort Pickens.

The retreating Confederate forces demolished almost everything standing at the yard. Although rebuilding by Federal forces began soon after, with peace came a lull that again brought a slack tide to construction. But tides change, like peacetime and war. War with Spain over Cuba and Puerto Rico brought an increased importance to Pensacola Navy Yard. New construction erupted, including new technology such as wireless stations and electric power plants. Along with the Spanish-American War, the construction of the Panama Canal gave the Navy yard a lease on life, if only temporary.

In 1906, the yard fell victim to the whims of Mother Nature. The area was hit by a hurricane, which, with winds measured at 105 miles per hour, was called "the most terrific storm in the history of Pensacola" in 170 years by the *Monthly Weather Review*. The *Review* put the amount of damage to the Navy yard at over $1 million; almost every building in the yard was either severely damaged or destroyed. The storm's effects, along with the presence of shipyards closer to Washington, such as Norfolk and Baltimore, lessened the importance of Pensacola Navy Yard, and it was officially closed in 1911.

One

THE BEGINNING TO 1940

The same year that saw the closure of Pensacola Navy Yard, 1911, was when Navy lieutenant Theodore G. Ellyson made the first aircraft landing on the USS *Pennsylvania* in San Diego. The year before, Ellyson had successfully taken off from the USS *Birmingham* in Hampton Roads. Glenn Curtiss had established his flying school in San Diego, and many experiments with naval aviation were being conducted at Annapolis. Now, Congress was persuaded to include a naval air service in the budget. But where to build a naval aeronautic station? Secretary of the Navy Josephus Daniels toured the abandoned yard in the spring of 1913 and sent his assistant secretary, Franklin D. Roosevelt, in the fall of that year. It seemed that an unused naval facility in the warmer southern climate presented itself as the ideal solution, and on February 2, 1914, the headline on the *Pensacola Journal* read: "World's First Naval Aeronautical Station Opens Here Today with Flights over Pensacola Bay and the Gulf by Ensign Chevalier and Other Aviators."

The USS *Mississippi* (BB-23) sailed from Annapolis on January 20, 1914, bringing the aviation unit, which was made up of 9 officers and 23 men. They brought with them seven aircraft and enough equipment to house and repair the equipment. The aviation unit had previously spent eight weeks at a camp on Fisherman's Point, Guantanamo Bay, Cuba, in 1913. The Marines arrived from Philadelphia in February on the transport *Hancock*. They had operated a flying camp on Culebra, off the coast of Puerto Rico. The unit lost no time in setting up tents, building wooden ramps, and establishing new records for height and endurance, while additional detachments of aviators and their planes continued to stream into Pensacola.

Although this photograph shows the aftermath of a 1916 hurricane, it can be imagined that this is what Lt. Comdr. Henry C. Mustin was describing on his arrival in Pensacola: "the beach was in a fearful state with wreckage of all kinds, bricks, stones and old railroad iron. . . . Tomorrow will start the ship's crew clearing away wreckage and building runways. This morning it looked like the ruins of a prehistoric city."

The *Pensacola Journal* reported that "the transport *Prairie*, with eight hundred men of the second reserve division of the Marine Corps aboard, arrived in Pensacola shortly after the noon hour . . . commanded by Col. Lejeune. Orders may arrive this morning for the marines to take up quarters at the navy yard. The buildings there are not all ready for occupancy by the men, but as one of the high officials stated last night, the Marines are supposed to be equipped fully and can make themselves comfortable at any point, so it will not be necessary for the buildings to be ready as they can live in the open until such time that the barracks are finished." Perhaps this undated photograph shows the encampment where they decided to "make themselves comfortable."

This aerial photograph of Naval Aeronautic Station was taken in 1915. The Navy yard is in the front quadrant, with the village of Woolsey to the right and Warrington on the left. The two villages were built on the naval reserve in the early 19th century to house civilian workers, as the reserve was, at that time, a six-mile commute from the town of Pensacola.

A Curtiss flying boat cruises over Pensacola Bay in front of the newly created aeronautic station. Note the USS Mississippi (BB-23) on the right. This photograph would have been taken between February 1914, the date of the station's creation, and July 1914, when the Mississippi was sold to Greece. The armored cruiser USS North Carolina (ACR-12) became the new station ship. (Photograph from Maritime Quest.)

The list of the naval aviators in this photograph, taken on March 6, 1914, reads like a map of the streets at NAS Pensacola. Shown here are, from left to right, Godfrey Chevalier, Richard C. Saufley, Patrick N.L. Bellinger, John H. Towers, William McIlvaine (USMC), M.L. Stolz, Bernard L. "Banney" Smith (USMC), and Victor D. Herbster.

Posing in front of a Curtiss flying boat is the first class of naval aviators in Pensacola. Many had already flown at Annapolis. Shown are, from left to right, (first row) Richard Saufley, Patrick N.L. Bellinger, Kenneth Whiting, Henry C. Mustin, Albert Read, Earle Johnson, Alfred A. Cunningham (USMC), Francis T. Evans (USMC), and John Haas; (second row) Robert Paunack, Earl Spencer, Harold Bartlett, ? Edwards, Clarence Bronson, William Corry, Joseph Pugh Norfleet, Edward McDonnell, and Harold W. Scofield. Some of these men, such as Saufley, Bronson, Haas, and Corry, gave their lives to naval aviation. Others, such as Paunack, Spencer, Norfleet, Bellinger, Whiting, and Read, distinguished themselves as career officers in the Navy.

Shown here are some early pioneers of the "July class, 1915, U.S. Navy Aeronautic Station." This was a period of experimentation and development in naval aviation, from command structure to aircraft structure. In his comprehensive work *United States Naval Aviation, 1910–1995*, Roy A. Grossnick explains that "based on recommendations received from the Naval Aeronautic Station [in 1915] . . . the Director of Naval Aeronautics established requirements for thirteen instruments to be installed in service aeroplanes: air speed meter, incidence indicator, tachometer, skidding and sideslip indicator, altitude barometer, oil gauge, fuel gauge, compass, course and distance indicator, magazine camera, binoculars, clock, and sextant."

In this informal photograph of early aviators, the men, listed with their naval aviator designation numbers, are, from left to right, ? Edwards, Clarence Bronson (15), Godfrey Chevalier (7), William McIlvaine (12), and Kenneth Whiting (16). Bronson was killed in Maryland in 1916 while testing bombing from aircraft. Chevalier earned the Distinguished Service Medal in World War I before his death in a plane crash in Virginia in 1922. Whiting earned the Navy Cross for his service in World War I and remained in naval aviation through World War II.

The fledgling aeronautical unit in Pensacola was called to action in April 1914 during the crisis with Mexico known as the Tampico Affair. Two ships, the USS *Birmingham* (CL-2) and *Mississippi*, transported two detachments along with five aircraft, including the AH-3 hydroaeroplane and the AB-3 flying boat.

Lieutenant Bellinger (far right) conducted several flights over Veracruz, making observations in support of ground troops and photographing the area with Ens. Walter D. Lamont. On May 2, he and Lieutenant Saufley achieved another first for naval aviation, as they took enemy fire while on a reconnaissance mission.

Other naval vessels visited Pensacola in the early years. This may be the *Narwhal* (D-1), a submarine commissioned in 1909 that patrolled the waters of the Caribbean and the Gulf of Mexico from 1913 to 1914.

Flying 600 feet above Pensacola Bay are "Smike" Connell (left) and Jerry Plumb. Lt. Byron J. Connell went on to attempt the first nonstop flight to Hawaii in 1925, with Comdr. John Rodgers. Unfortunately, he ran out of fuel just shy of completing the voyage.

H Bronson.

Wind about 5 mph.
Elevator considerably "down."

Early efforts at launching a plane with a catapult were made on a converted coal barge anchored to the dock. This 1915 photograph shows Lt. Clarence Bronson being launched in a Curtiss AB flying boat.

One can only imagine the feeling of being launched into the air in 1915 in such a precarious-looking craft. As one early aviator, Capt. Edward "Ned" Wenz, put it, "It was exhilarating."

Not all flights were exhilarating. Many ended up in Pensacola Bay. This Curtiss pusher is being prepared by two aviators to be hoisted out of the water by a crane. Lieutenant Chevalier is at left on the plane. Many pilots were killed not by the impact, but by being struck in the back of the head by the "pusher" motor. Pusher aircraft were soon removed from service.

The cost of experimental aviation was high: a few accidents resulted in the death of the pilot. Here, a funeral procession crosses Palafox Place on Garden Street, between 1914 and 1917. The *Pensacola Journal* lists five aviators killed in Pensacola during that time: Lt. J.H. Murray, February 1914; Lt. M.L. Stolz, May 1915; Lt. James V. Rockwell, May 1916; Lt. Richard C. Saufley, June 1916, and Ens. Dean R. Van Kirk, May 1917. Another difficult year was 1918, particularly in September and October, when Ensigns Louis J. Bergen, Joy C. Bournique, Thomas C. McCarthy, and Carl O. Peterson were killed in seaplane and airplane accidents.

Often, the cause of a pilot's death was drowning, the result of the pilot becoming entangled in the wreckage. Aided by an observer in a tower onshore, the crash sled was an attempt to minimize fatalities by getting to the site of a downed pilot quickly. It patrolled a regular course, and trainee pilots were required to fly the same course in the air so the crash sled would be nearby if they did fall. The sled came to be known as "the solo coffin," and it must have been a sobering reminder to pilots as they saw it cruising around below them.

Sometimes, towing a downed wreck through the water did more damage to the aircraft than the impact with the water. A solution to this problem came in the form of the *Mary Ann*, a wrecking barge designed to hoist a plane out of the water. This craft was also used to get planes on the improvised coal barges in the early days.

This early photograph is labeled "Kelly Upton stubs his toe *Mary Ann* to the rescue Pensacola." The *Mary Ann*, so named for an officer's daughter at the aeronautic station, was described by the *Pensacola Journal* in 1916 as having "two pontoons, connected by trusses and decking, on which a derrick, with a five-ton capacity and hoisting speed of five feet a minute is installed."

Another impediment to early aviation was the weather. This photograph shows the damage to the wet basin area from the many tropical storms of the 1916 season, which pulled down much of the new construction.

Lt. Comdr. Henry Mustin makes history, piloting a Curtiss AB-2 in the first catapult launch from a ship underway, on November 5, 1915. The ship was the armored cruiser USS *North Carolina* (ACR-12).

Capt. Francis Evans (USMC) added to Pensacola's list of firsts on February 13, 1917, when he performed a loop in a Curtiss N-9 float plane on his fourth attempt. No one believed that the feat could be done. What Evans learned about stall and spin recovery is still being taught to pilots today.

Another first at NAS Pensacola was the initial Coast Guard aviation group, shown here in front of a Curtiss N-9 on March 22, 1917. The caption lists the members as follows: "Left to right: C.T. Thrun, Mas. At Arms; J.F. Powers, Oiler 1st Cl.; Geo. Ott, Ship's Writer; C. Griffin, Mas. At Arms; John Wicks, Surfman; Robert Donohue, 3rd Lt.; C.E. Sugden, 2nd Lt. (Eng.); E.A. Coffin, 2nd Lt.;S.V. Parker, 1st Lt.; P.B. Eaton, 2nd Lt. (Eng.); E.F. Stone, 3rd Lt.; Ora Young, Surfman No. 1; W.R. Malew, Coxswain; J. Myers, Surfman; J. Medusky, Asst. Mas. At Arms; R.F. Gillis, Signal Q.M.; W.S. Anderson, Surfman; L.M. Melka, Signal Q."

It was recognized early on that weather conditions affected flights, thus the first Aerography School was established on May 8, 1917. The first aerographic flight with meteorological instruments was made in April 1918 by Lt. William F. Reed. The original logbook of the Meteorological Department can be found at the National Naval Aviation Museum (NNAM).

Other early experiments in aviation included free and tethered balloons. This photograph is from one of many scrapbooks in the Emil Buehler Library at the NNAM. The caption reads, "The water borne dirigible shown entering the floating hangar, April 1917."

This photograph shows the immense dirigible hangar secured at the wet basin. This hangar could be towed out to sea and placed to accommodate wind direction. The dimensions of the early dirigibles were as much as 175 feet long and 35 feet wide.

This photograph's caption reads, "Observation kite and 'Eagle Boat' pictured at NAS Pensacola, 1921." The observation kites accompanied destroyers as submarine hunters, but were not always beneficial. Many times, they would give away their convoy's position. Henry Ford was instrumental in the design and construction of the Eagle Boats, used to patrol US waters at the end of World War I.

B-class airships, like the one shown here, first arrived in August 1917. They were built in Akron, Ohio, and used mostly for training and for patrolling the coasts during World War I. The B-class was a great improvement over the DN-1 (later labeled A-class by the Navy), which quite literally never got off the ground.

Behind the glamour of flying planes and airships, the daily life of sailors and marines marched on. This photograph, dated August 8, 1918, shows an encampment on Santa Rosa Island, a mile across Pensacola Bay from NAS Pensacola.

Another aspect in the life of a sailor is inspection, like this one, held on July 27, 1918, probably at morning quarters. The ritual of quarters is all about muster, inspection, and instruction: mustering the sailors, inspecting uniforms and grooming, and giving marching orders. Quarters is also used to give out awards and commendations. While the regular Navy still holds quarters, the fast-paced world of naval aviation today holds turn over, maintenance meeting, and passdown. Muster is done by individual work centers and reported up, and inspection is done as needed. Instruction comes from maintenance meetings with all supervisors and is then passed down to the sailors.

In the tradition of improvisation, these sailors made use of the newly installed catapult track on the USS *North Carolina*. Someone's family at home read this inscription: "On our way to the 'war gone.' Our reading room—on the boat deck under the blue sky and on the blue sea."

In 1915, there was no accommodation for landing an aircraft aboard any Navy ship, so this Curtiss AB-3 flying boat is being hoisted aboard the *North Carolina*. It was certainly a complicated procedure.

Once hoisted aboard, the aircraft was strapped to this carriage. To get the plane aloft in a short amount of distance, a complicated system of compressed air, cables, and pulleys were used to increase the force by seven times. When the plane reached the end of the track, the straps released and the plane was catapulted into the blue.

Seen here is the catapult track from above as it winds its way to the bow. The *North Carolina* was commissioned in 1908 in Norfolk, Virginia, and had quite a career, including returning the bodies of sailors killed in the 1898 explosion aboard the USS *Maine* (ACR-1), which were then interred at Arlington National Cemetery. The *North Carolina* was sold for scrap in 1930.

Much as with today's carrier flight-deck operations, it takes the teamwork of an entire crew to prepare an aircraft for launch. Clearly, the launching of this Curtiss Model F (later AB-3) was enough of a novelty to attract the attention of all, including the civilian just to the left of the track.

An empty car at the end of the steel track suggests a successful launch. Note the construction of the track and its height above the deck of the *Carolina*. An article in the *Pensacola Journal* in May 1917 describes the launching system: "The apparatus is composed of a track, a compressed-air cylinder, a car to run on the track, and a cable connected with the piston of the air cylinder at one end and with the car at the other end. Seated in his aircraft, the pilot starts his motor, and with the propeller running at full speed, the compressed air is released into the cylinder, which pulls the cable and the seaplane rapidly along the track. Tests show that in a distance of 108 feet of deck space the airplane thus attains a velocity of 45 miles an hour. A tripping device at the end of the track releases the cable and the plane, with motor spinning, shoots away from the stern of the ship and is off."

In a 1914 interview with a *Pensacola Journal* writer, "Captain Mustin stated . . . that he was preparing to open the machine shop at the yard in a very few days. The aeroplanes will be repaired here and as soon as possible new aircraft will be built at the yard, the machinery there being capable of manufacturing parts for almost any kind of water or air craft." The following photographs, taken in 1919, show the details of the machinery division facilities, including this "carburetor shop."

This is the "motor erecting shop," housing the tools and equipment used to maintain and repair the aircraft's engines. The new aeronautic station was very self-contained, as it was, at the time, fairly isolated from supplies and resources.

The "propeller shop" was always ready to cope with the mishaps of pilot training. One of the old propellers can be seen in the Emil Buehler Library at the National Naval Aviation Museum.

The "foundry," built in 1882, is described in a report to the Bureau of Yards and Docks in 1905 as being "entirely too small," and funds were requested for improvements.

Experimentation and research go hand in hand. This Curtiss N-9H was built using research on wind tunnels conducted at the Massachusetts Institute of Technology. The N-9H was the first Navy aircraft built from this research.

Another plane employed by the Navy was the Burgess-Dunne AH-7, around 1915. This tail-less, swept-wing innovator was camouflaged with green and lavender. It was built in Marblehead, Massachusetts, and featured a 70-horsepower Renault engine.

This de Havilland two-seater biplane, in service from 1918 to 1925, has been outfitted with a compartment for a stretcher and marked with the Red Cross symbol. It is parked in front of one of the steel hangars that lined the water's edge.

Later in the 1930s, this N3N-6, called "The Yellow Peril," represented the only model of aircraft built by the US Navy. Produced at the Naval Aircraft Factory at the Philadelphia shipyard, the N3N was in service with the Navy until retired in 1961.

When the Navy first arrived, wooden ramps with fabric tents were built to hold the aircraft. Teams of sailors were used to pull the seaplanes up the ramps and into the tents.

By 1917, the wooden ramps were replaced with cement, and the tents were replaced with steel hangars. Small, flimsy tents were inadequate to house the larger aircraft being purchased by the Navy at the time of US entry into World War I, so several steel hangars were built to house the craft and provide space for maintenance and repairs.

Shown here is a typical uniform of a naval aviator in 1917. Capt. Edward "Ned" Wenz was Naval Aviator No. 224. He retired from the Navy in 1953. (Courtesy of the Wenz family.)

A group of Yeomanettes, as they were commonly called, stands in front of a building at NAS Pensacola about 1917. Officially, they were labeled on paper as "Yeoman (f)" so they wouldn't inadvertently be sent to sea. Their most common function was that of office staff, freeing up their male counterparts for more "manly" duties. The first Yeomanettes enlisted in March 1917, and the rank remained on the books until 1921. An excellent book on the Yeomanette experience is *The First, the Few, the Forgotten: Navy and Marine Corps Women in World War I* by Jean Ebbert and Marie-Beth Hall.

According to the caption, this is "Personnel inspection/B. Green 10 Nov 1918." This photograph was taken the day before Armistice Day and a month after Germany asked for a general armistice. Many of these sailors were likely being inspected for the last time and looking forward to going home.

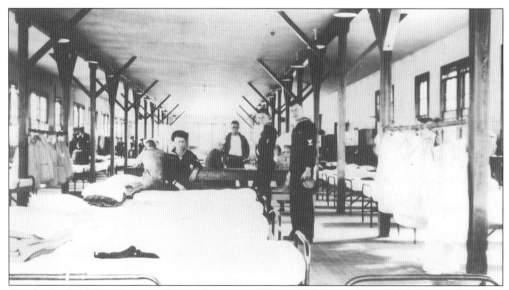

On the back of this photograph, a sailor wrote: "Circa 1918 Barracks at NAS Pensacola. Close quarters in the barracks during wartime." In 1917, the aeronautic station held only 163 enlisted men. By the end of the war, naval aviation saw the increase of enlisted sailors to over 5,000.

Sailors pose with a "1920 seaplane with Curtiss Motor," as written on the photograph. Researchers at the Buehler Library believe this is a Wright-Martin R and state, "In September of 1916 Glenn Martin Co and General Aeronautical Corporation formed the Wright-Martin Co joined in a short venture to produce Aircraft. They built the Model R and V, both land based tractor type but did have several R models with floats."

This photograph features the gunnery class of January 24, 1919, "leaning up against a camouflaged hangar." During World War I, most of the larger buildings at the aeronautic station were given the camouflage paint scheme. Note the Marine Corps officer in the third row, center. He is most likely their instructor.

In this undated photograph, radiomen work in the radio repair shop. NAS Pensacola was a leader in developing radio use in aviation. According to the Naval History and Heritage Command, the "Chief of Naval Operations requested that two small seaplanes and one pilot be detailed for duty in connection with radio experimentation at Pensacola" in May 1917.

After Armistice Day in 1918, activity at the base slowed, but experiments and training continued. By 1922, outlying and auxiliary fields were built to create longer runways. Many of the runways, however, had already become obsolete by World War II because they were too small to handle the heavy aircraft of the time. Some runways were little more than an old cow pasture, but a few were eventually paved and included buildings. Even fewer are still active installations, such as Corry Station and Saufley Field. In 20 years, over a dozen fields of varying size were built. Detailed information can be found on Paul Freeman's excellent website, www.airfields-freeman.com. This photograph shows "approximately 40 wave department (Curtiss) JN-4H Jennys in foregrounds."

In this undated photograph, a Loening OL-1 takes off from the USS *Langley* (CV-1), the Navy's first aircraft carrier. The *Langley* was renamed after being repurposed from the collier *Jupiter* at Norfolk in 1920. The Loening first flew in 1923.

This aerial photograph of the *Langley* shows an assortment of aircraft on deck. The *Langley* was in service until being scuttled in the Pacific in 1942 after a Japanese bombing. The attack resulted in 16 crew members being lost. They are considered still on duty.

Sailors found other uses for radio than strictly communication. In this 1922 photograph, they offer a "concert" in their off-hours for the lofty price of 10¢.

Another hurricane struck in 1926. The greatest damage occurred in the Miami area, but winds and flooding due to the storm surge caused considerable damage to the station, leaving the sailors with a huge cleanup job.

A Reserve Officers training group poses in 1923, three years before the NROTC was officially established. In 1926, six units were established in colleges and universities throughout the nation. In 1932, the Marine Corps joined the program. Then, as now, the training program provides opportunities for citizens to gain an education that will allow them to become Navy and Marine Corps officers.

Shown here are the Navy and Marine Corps "Graduates of Class XIX, 2 February 1924, Naval Air Station, Pensacola, Fla." By this time, the *Langley* was a part of the fleet, and instruction would have included carrier takeoffs and landings.

This photograph is captioned, "Officers and Enlisted Personnel Squadron 'V' Pensacola, FLA Naval Air Station 1927." The men pose together in front of a hangar and an early Curtiss seaplane.

Posing in front of its Curtiss F6C Hawk is "Combat Class #13 Corry Field, Pensacola, Fla. 17 Jan 1930." Shown are, from left to right, (seated) Jack Monroe, ? Bayler, Bob Armstrong, and Benj Stratton; (standing) Bill Harris, Jimmy Hatch, Jack Tracy, Jack Griffin, Clif Cooper, Stan Dunlop, Ace Conrad, Howell Dyson, C.R. Shanahan, and Ernie Rust.

Shown in this undated photograph of aviators and their mascot are, from left to right, (first row) Haynes, Winkle, Ball, Ahlson, Smith, Ashwell, Hall, Grover, Venables, and Hess; (second row) Paton, Harrington, Newling, Young, McGinnis, Langan, Coney, Noon, Colt, Putnam, Blackmare, and Barkus.

This photograph is labeled as follows: "Naval Reserve Aviation Division V N 5 R042, 9 Sep 1933." The plane is a Curtiss F8C-4 Helldiver, a fighter-bomber that saw service in the Navy in the 1920s. They were later used as observation planes.

Gathered on May 10, 1934, in front of a Vought OS2U, are, from left to right, (first row) Massey, Lawrence, Lytle, Farquarson, Gertner, Hatton, Raborn, Crane, Patten, O'Neill, Greene, Metsger, Lucas, and Kelsey; (second row) Ch. Croft, Henry, Cooper, Ramsey, Bruner, and Zeigler; (third row) Spencer, Donaldson, Donohue, Beauparlant, R. Brown, Bronson, Winsett, Murray, Lewis, Gwinne, Lovern, MacDonald, Reynolds, Gowan, Fox, Scott, C.E. Pierce, Lee, Wilson, Meilozanski, Lamb, Wilezewski, Simmons, Singletary, W.R. Brown, Finlayson, Johannson, Binder, Williams, Ascroft, Nine, Bryant, Cusick, and Gurholt.

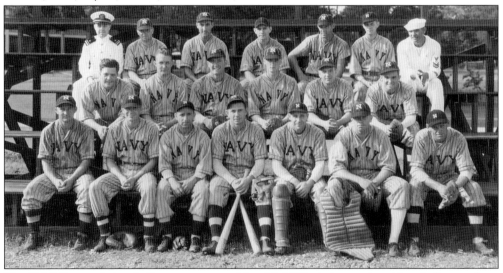

The NAS Pensacola baseball team poses on May 26, 1938. Aviators played baseball as early as March 1917, when they played the team of the USS *Bushnell*. The tradition of baseball and the military dates back to the Civil War, and many great baseball stars played on Navy and Marine Corps teams, including Ted Williams, Joe DiMaggio, and Ty Cobb.

Two

WORLD WAR II

All the experimentation and training showed itself to be fortuitous when the United States was drawn into World War II. One step in the preparation process had been the creation of the cadet training program in 1935. The Navy only had 1,277 planes in 1939. But President Roosevelt, who had always been a proponent of the Navy, signed the Two-Ocean Act in 1940, which authorized 15,000 aircraft. Soon, an underused station that had feared another closure and that trained only about 800 pilots a month became the premier training base for naval aviators, sending out 2,500 aviators monthly. According to Robert Lawson and Barrett Tillman's excellent book, *World War II U.S. Navy Air Combat*, the Navy had nearly 8,000 student pilots at the beginning of the war. Within a year, there were four times that number in primary flight training alone. To accommodate the increase, "primary training was moved from Corpus Christi and Pensacola to Naval Reserve bases across the country." Intermediate training meant "fourteen weeks in Corpus Christi and Pensacola," where the pilots flew Boeing-Stearman N2s, then were carrier-qualified at Great Lakes San Diego. Pensacola housed five squadrons, for training in sea and land planes. Also, many firsts were accomplished, such as the use of WAVES (Women Appointed for Volunteer Emergency Service) as gunnery teachers, landing signal officers, and photographers. Pilots from Allied nations such as Britain and France were trained at the station. Additionally, two pilots from NAS Pensacola, Lts. Stephen Jurika and Henry L. Miller, helped train the members of Doolittle's Raiders at Eglin Field in preparation to perform carrier takeoffs in B-25 Mitchell bombers, according to the NNAM.

Vought SBU-1s are parked in the Squadron 3 Airship area of Chevalier Field around 1941. The Vought Corporation, based at that time in Dallas, Texas, built carrier-based aircraft for the Navy during World War II. The airship hangar is long gone, replaced by the new Chevalier Hall; but the brick hangars on the right remain today.

Two years later, Chevalier Field was packed with Douglas R4Ds (the Navy's designation for the Army C-47 and civilian DC-3), Beech SNBs, and SNJ Texans.

An American sailor trains French sailors in an SNJ Texan. Many Allied pilots from several nations, including France and Britain, trained at NAS Pensacola during World War II.

Here, SNJ Texans are lined up at Corry Station, about five miles north of NAS Pensacola. The Texan was an intermediate trainer during a pilot's course of instruction.

About three miles farther north from Corry Station is Saufley Auxiliary Field, seen in this view looking south on September 4, 1943. Note the aircraft lined up on the runways. After World War II, activities were scaled back, but in 1968, Saufley became a naval air station once again to support the need for trained pilots during the Vietnam War.

Here, WAVES tend to innumerable SNJ Texans at Whiting Auxiliary Field, located near Milton, Florida, about 40 miles northeast from NAS Pensacola.

Barin Field was another auxiliary airfield, created when Foley Field was leased by the Navy in 1942. It was located about 30 miles west of NAS Pensacola, near Foley, Alabama.

An F6F Hellcat is seen after a bad landing at Barin Field, which earned the nickname "Bloody Barin." As the pressure to train pilots in a shorter amount of time increased, so did the number of training accidents. No airfield had a higher incidence of fatalities than Barin; in its first two years, 40 fatalities occurred.

"An instructor demonstrates maneuvers to a student in the backseat of an N3N-1." Almost 1,000 of the "Yellow Peril" were built by the Naval Aircraft Factory in Philadelphia from 1935 to 1941. The aircraft was used by the Navy until 1961. An excellent example of the N3N can be seen at the National Naval Aviation Museum.

Sailors pose in 1943 in front of a North American SNJ Texan, a trainer during World War II.

Marine Aviator Joe Foss flies a Texan over Pensacola during World War II. Foss earned the Medal of Honor for his actions in the Pacific theater, particularly at Guadalcanal. He died in 2003, two years after establishing the Joe Foss Institute in Scottsdale, Arizona, to promote patriotism and public service.

Aviation cadets take a break from instruction in the mid-1930s. Some cadets used their break time to get in some more studying. A few others caught up on much-needed sleep, as aviation training was quite grueling.

In this photograph, nine Texans fly in right-echelon formation, with the beautiful sandy beaches of Pensacola below. The word *echelon* comes from the French *échelon*, "ladder rung."

In this view of the Texans, they are flying in left-echelon formation. Echelons have been used in military formations since ancient times, because they give a greater depth of vision.

This 1940s photograph shows a pilot about to experience "Dilbert the Dunker." A pilot would be strapped into the cockpit of an SNJ Texan perched atop a 45-degree rail, high above a deep pool.

The next step of the "dunker" exercise involved the cockpit being released and upended on impact with the water. The pilot had to exit the cockpit and swim away from the "craft" at an angle, because in many water crashes, the surface of the water is covered with burning oil and gas from the downed craft.

Pilots gather for a preflight check; this photograph was taken at Saufley Field during World War II. According to the board, the pilots were urged to do the following: "Pre-flight your plane—carefully. All pilots check out and in. Get your plane number, move away from the board. Use extra caution taxiing."

For this photograph, dated December 18, 1943, the caption reads, "Engine Storage Bldg. #693. Engine stored on pipe stands exposed to the air. At that time the silica gel method was used to detect the presence of moisture. In June 1944, it was directed that all engines be boxed."

This photograph is labeled "Parachute loft personnel Mar 1944." Note the WAVES in the back two rows. The personnel shown here are, from left to right, (first row) Williams, Morris, Swank, Newton, Leopard, Domer, Russo, McDonald, Baron, Keen, Stack, and Hensel; (second row) Coughlin, Ordille, Kleinschmidt, Mahon, Sharp, Schoenborn, Fillinjim, Chief Hobbs, Lt. Stillwell, Chief Wright, Gilroy, McDonald, Kyle, Wash, and Feld; (third row) Medford, Morse, Aackre, Adamaitis, Nowell, Gegan, Welch, Miller, Zdonowics, Hogan, and Howells; (fourth row) Nesbit, Jorgenson, Rockler, Joy, White, Moore, Dasbach, Rusiecki, Wooley, DeVore, and Ewing.

WAVES also taught gunnery during World War II. The members of this group are, from left to right, (first row) Tennessee (Phyllis Roberts), Jean Thurck, Evelyn Marsh, Mr. Perry, Anne Holder, Jean Bullett, and Jig Shaft; (second row) Betty Hayes, Helen Shay, Winnie Donellan (?), Jani Wherle, Adelaide Silver, and Pat McNutt; (third row) Jean Wear, Jo Jenouski (Sachs), Betty Brackett Robinson, Peg Pritchard, Jennie Leied, Eva Christensen, Martha Fell, Betty Rucker, Anne O'Brien, and Betty Cottrell; (fourth row) Jindie Bys (?), Jo Puckett, Helen Studlieu, Rokie Graves, Flossie Bruss, and Hazel Smith; (fifth row) June Osbourne, Claire Calloway, Shirley Dickstew, Yvonne Forbes, Helen Dolan, Mary Burmingham, Marion Browne, Pat Connolly, Alice Robinson, Hazel Brown, and Sally Moore. The names of Martha Turner, Connie Alzonzi, Carol Adams, Dorothy Sipe, Ethel Kimzie (?), and Edna Arut are also listed. (Photograph and list of names courtesy of Betty H. Carter Women Veterans Historical Project, University of North Carolina at Greensboro.)

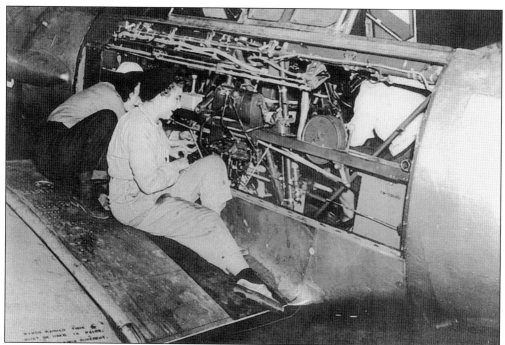

WAVES were also employed to keep the planes in repair. These WAVES work on an SNJ Texan in a hangar. (Courtesy of National Archives.)

WAVES maintain this Texan outside in the Florida sunshine. The Naval History and Heritage Command records that "8,000 female officers and some ten times that many enlisted WAVES, about 2-1/2 percent of the Navy's total strength," worked alongside male sailors during World War II. (Courtesy of National Archives.)

World War II may have ended in 1945, but threats to peace continued to embroil many areas of the world, and the Cold War was just beginning. Naval aviators, both active and reserve duty, continued to train. This photograph shows a formation of F6F Grumman Hellcats of Naval Air Reserve Training Unit (NARTU) CVG55/83 in August 1948.

These F6F Hellcats of NARTU CVG 55/83 train on the flight deck of the USS *Wright* (CVL-49) in the Gulf of Mexico. The *Wright* was later converted to CC-2, a "Doomsday Ship." Along with her sister ship, USS *Northampton* (CC-1), the *Wright* kept watch on the East Coast, prepared to become a seaborne White House or Pentagon in case of nuclear attack.

These F6Fs are in echelon formation on March 28, 1951. A beautifully restored example of the Hellcat can be found at the National Naval Aviation Museum in Pensacola. According to the museum's website, the plane "represents a class of carrier-based fighter aircraft that was credited with shooting down more enemy aircraft than any other type of aircraft during World War II."

A North American T-28 Trojan is seen here, with Whiting Field as a backdrop. The Trojan, first flown by the Navy in 1949, was specifically designed to help pilots make the shift to jet airplanes. Two models were built, B and C, with C designed for carrier flying.

Commissioned in 1934, the USS *Ranger* (CV-4) was the first carrier built for that purpose from the keel up. But by World War II, the ship was becoming obsolescent. She was sent to the Atlantic, where she became the only US carrier to participate in attacks against Germany, off the coasts of Norway and Africa.

Here, the *Ranger* is underway in the Gulf of Mexico. Note the presence of three funnels on the starboard side near the stern. The aircraft are thought to be TBD Devastators.

To facilitate launching and traps, the three funnels of the *Ranger* were turned on their sides during flight operations.

The *Ranger* was decommissioned in October 1946, but not before spending a brief time in Pensacola. The biplanes on the flight deck may be the Great Lakes BG-1.

Three

THE JET AGE AND THE COLD WAR

As the world entered the age of jet-powered aircraft, the Navy was not going to be left behind. But, like their predecessors a generation earlier in their Curtiss flying boats, jet pilots were entering uncharted waters. Many experiments cost the lives of good men, and this was unacceptable. In answer to the problem, Advanced Training Unit (ATU) 206 was inaugurated at NAS Pensacola in 1955. The unit used 72 Grumman F9F-2 Cougars to train new pilots. A former naval aviator affectionately described the Cougar as "the 'lead sled' because it was underpowered, built like a tank to withstand the rigors of carrier operations. It was, however, an aircraft that was forgiving to pilot mistakes." It is often described as the Navy's first "successful" carrier jet, and for good reason. Many years of experimentation and accident had preceded the F9F's flights in the Pensacola skies.

The growth of aviation brought other significant changes to NAS Pensacola. The runways of Chevalier Field were no longer capable of handling the powerful jets, which needed longer runways. Forrest Sherman Field was built in 1954, and Chevalier Field became a training site for helicopters. Adm. Forrest Sherman was a Naval Academy graduate, a naval aviator during World War II, and he rose to become, at that time, the youngest chief of naval operations, in 1949. Sherman Field, according to NAS Pensacola, with its "8,000 foot runway, is the standard military length for a Navy and Marine Corps facility. It has the capability of launching and recovering every type aircraft within the US Military arsenal, even C17s and B52 Bombers, and can also handle many civilian aircraft including the new 787 Boeing Dreamliner."

This late-1950s photograph illustrates the transition taking place at the time. Shown arrayed at Chevalier Field are the following craft: the Lockheed T2V Sea Star (far right), which entered service in 1957; the Beechcraft T-34 Mentor (foreground), which began production in 1955; and the North American T-28 Trojan (center), which first flew in 1949.

The carrier *Monterey* (CVL-26) served as a training ship for carrier qualifications in the 1950s. Here, a F6F Hellcat from Barin Field prepares to be launched from the flight deck.

A pilot trainee works through the syllabus, or flight instruction program. The syllabus is a defined progression of instruction, from primary through basic to advanced flight training, including carrier qualification.

Here is another illustration of the transition from propellers to jet engines. This F6F Hellcat sits in front of the warning "Beware of Jet Blast" on the flight deck of the USS *Lexington* (CV-16).

This aerial photograph of Naval Air Station Pensacola, dated February 27, 1963, shows Sherman Field on the left. Note Chevalier Field on the right. Today, it is home to Chevalier Hall, which houses the Naval Air Technical Training Center (NATTC).

A Lockheed T2V Sea Star glides over the newly built Sherman Field. The Sea Star was an ideal carrier training aircraft, as it had lower takeoff and landing speeds.

A formation of Sea Stars soars over the USS *Antietam* (CV-36). The *Antietam* was assigned to NAS Pensacola in 1957 as a training ship, but Pensacola Bay was not deep enough. Until dredging made it possible for the *Antietam* to come home in 1959, she sailed out of Naval Station Mayport. The carrier remained at Pensacola until she was relieved by the USS *Lexington*. It was from the flight deck of the *Antietam* that the *StratoLab V* flew its historic record-setting high-altitude flight on May 4, 1961, in the Gulf of Mexico.

Another formation of Sea Stars flies over NAS Pensacola. Chevalier Field is in the upper right. Many of the buildings in this photograph are no longer standing.

A T-28 Trojan flies over the Gulf of Mexico, with Santa Rosa Island in the background. Not only was Pensacola an ideal place to establish an aeronautic station because of its mild climate, but the beauty of the sparkling gulf and sandy white beaches is still appreciated by sailors and their families today.

A sailor installs an aerial camera in a Grumman F9F-6P Cougar. Cameras were installed in 60 F9F-6Ps, which were specifically designed to carry cameras instead of cannon. The aircraft had forward and aft bays. Researchers at the Emil Buehler Library at the NNAM report that "the forward bay accepted the CA-8 and K-17 [camera] and the aft bay accepted the K-17, K-17C, and S7-5 (Somme) cameras with various focal length lenses."

The caption for this photograph reads: "5 Jun 1959 Air-to-air view of a F9F-8P Cougar Jet Photo Reconnaissance aircraft from the Naval Air Technical Training Unit (The Naval Schools of Photography) in which Naval Aviators received training as Photo-Recon pilots prior to joining fleet photo squadrons."

This photograph is captioned: "A pre-1965 photo shows NAAS Saufley Field, looking south." Saufley Field, still an active naval installation, is named for Lt. (jg) Richard C. Saufley, who was killed over Santa Rosa Island on June 9, 1916, while attempting to break his own endurance record.

In October 1962, the USS *Lexington* relieved the *Antietam* as the training carrier for NAS Pensacola. Here, the carrier is being escorted into its berth by a team of tugs. According to the Navy's website (www.navy.mil), the *Lexington* was the fifth Navy ship to bear that name. She was commissioned in 1943, and she earned 11 battle stars and the Presidential Unit Citation for action in World War II at Tarawa, Wake Islands, Kwajalein, the Philippine Sea, Mindanao, and Leyte Gulf. The ship's active service life was longer than any other Essex-class ship. In 1963, her status reverted to attack carrier during the Cuban missile crisis. The *Lexington* returned to duty as a training ship until 1991, when she was retired. Today, "Lady Lex" is a floating museum in Corpus Christi, Texas.

This bird's-eye view of the *Lexington* at her pier gives the viewer a sense of her sheer size. Chevalier Field is on the extreme right, and the old seaplane hangars are on the left. Most of the buildings in the center are gone, victims of hurricanes, except Building 38 at the wet basin, Building 16, and the smokestack of the power plant.

The "USS *Lexington* [is] underway on flight ops with North American T-28 Trojans parked on the forward flight deck." In a 1983 letter from the General Accounting Office (GAO) to the Appropriations Committee, the *Lexington*'s training is described: "The *Lexington* is homeported at Pensacola and performs its training mission in the Gulf of Mexico within 70 to 100 miles of the shore-based airfields. Thus, student pilots fly out to the carrier from the shore-based facility, perform the requisite number of takeoffs and landings, and then return to base. The *Lexington* can currently accommodate all the Navy's training aircraft and A-6 and A-7 combat aircraft (for fleet pilot training) but cannot accommodate other types of combat aircraft, such as the F-4, F-14, or F/A 18."

Written on the back of this photograph is "Gulf of Mexico—catapult crewmen aboard the carrier USS *Lexington* hook up a TF-9J Cougar advanced trainer aircraft for launch. The aircraft is piloted by Naval Aviation Cadet Gene L. Porter of Oak Harbor, Washington, who is undergoing carrier qualifications aboard the *Lexington* prior to graduation from the Navy-Marine Corps Aviation Cadet program."

"Carrier Ops Flight Deck Operations F9F-8 prepares to land aboard USS *Lexington*." The 1983 letter from GAO states that "the Navy would prefer not to use deployable carriers to train student pilots on a continuous basis because it believes this practice could adversely affect fleet operations, crew training and crew morale. Navy officials said that [the practice] would extend its at-sea time an average of three weeks a year."

A Sikorsky H-34 Choctaw from an HC squadron at Ellyson Field "flies in the SAR (Search and Rescue) pattern for rescue missions with the *USS Lexington*." Research at the Buehler Library suggests that the H-34 was used in the training command until 1969 and was replaced by the H-1. NAS Ellyson Field, used by the Navy from 1941 to 1979, was named to honor Comdr. Theodore Ellyson, the Navy's first aviator. Helicopter pilot training began in Pensacola when Helicopter Training Unit-1 moved to Ellyson Field in 1950 and remained there until Pensacola's civilian airport traffic created a conflict with Ellyson Field. In 1972, the command relocated to NAS Whiting Field.

After World War II, helicopter training was another important facet of NAS Pensacola and its outlying fields. This Bell HTL-7 hovers over Ellyson Field in 1957. According to the staff of the Buehler Library, the "HTL-7 was the first 2-seat cabin with dual controls. Delivered in 1958, they were designated as TH-13s in 1962." Additional helicopter training was also done at Chevalier Field and may have continued until the mid-1990s, as US Geological Survey aerial photographs show helicopters on the field in 1994.

From the NNAM: "TH-13 helicopters [are] pictured lined up at Naval Auxiliary Air Station (NAAS) Ellyson Field in 1966, forty-seven years ago, many of the flight students that flew them at that time destined for service in the expanding war in Vietnam."

The caption of this photograph states that the "last of the UH-19F helicopter, BC no. 8585, departs Ellyson Field for the last time on 24 Sep 1964." According to the chief of naval training (CNATRA), "Once they receive their Wings of Gold, Navy and Marine Corps helicopter pilots report to their respective FRS (Fleet Replacement Squadrons) for further training. The Navy also trains helicopter pilots for the Coast Guard and several allied nations." (Courtesy of CNATRA.)

A Beechcraft T-34 Mentor taxis at NAS Whiting Field with T-28B Trojans parked in the background. A former instructor at NAS Pensacola reported that "only T-28s were based at Whiting, and were used for basic training. There were three parts to the training syllabus, basic flight, instruments and formation." A plaque honoring the field's namesake reads: "Whiting Field, named in honor of Capt. Kenneth Whiting, U.S. Navy, Pioneer in Submarines and Aviation, Naval Aviator No. 16, Father of the Aircraft Carrier in our Navy, Died on Active Duty on April 24, 1943." During his service, Captain Whiting earned the Navy Cross and Decoration of the Legion of Honor.

This photograph shows two T-34s in the skies over Pensacola. The T-34 was the primary trainer for decades. They were based at Saufley Field, where primary training was conducted. As of 2013, it is being replaced by the T-6 Texan II.

In 1948, Jesse LeRoy Brown became the Navy's first African American aviator. According to the Naval History and Heritage Command, "he was assigned to Fighter Squadron 32, flying F4U-4 Corsair fighters" over Korea. He was shot down in December 1950 and posthumously awarded the Distinguished Flying Cross. The Navy named an escort ship in his honor in 1973. In July 2013, on the 60th anniversary of the Korean armistice, Brown's wingman, Medal of Honor winner Capt. Thomas Hudner, returned to North Korea in an attempt to locate Brown's remains, but was turned back because of severe weather conditions. (Courtesy of National Archives.)

Another first for naval aviation during this time was the first female aviators, in 1974. Pictured here are, from left to right, Lt. (jg) Barbara Allen, Ens. Jane M. Skiles, Lt. (jg) Judith A. Neuffer, and Ens. Kathleen L. McNary.

Posing with their helmets are, from left to right, Ens. Rosemary Conaster, Ens. Jane Skiles, Lt. (jg) Barbara Allen, and Lt. (jg) Judith Neuffer. According to an exhibit at NNAM, *From Typewriters to Strike Fighters: Women in Naval Aviation*, Allen was killed in a training accident in 1982. Conaster, Skiles, and Neuffer retired as captains. The exhibit is available online at www. navalaviationmuseum.org.

Female pilots were spared no quarter and trained exactly as did their male counterparts. Here, Lieutenant (jg) Neuffer prepares to experience "Dilbert the Dunker."

Survival training included "several days of wilderness skills training in remote areas of Eglin Air Force base," which is about 50 miles east of Pensacola. Eglin contains thousands of acres of forest, ideal for such training. According to the base fact sheet, "In March 1942, the base served as one of the sites for Lieutenant Colonel Jimmy Doolittle to prepare his B-25 crews for their raid against Tokyo."

Ensign Conaster inspects a landing gear strut during a preflight check. In checking the struts, the inner cylinder that compresses into the outer cylinder must be confirmed to be clean and nick-free, otherwise she might lose her seal and, hence, the effectiveness of that strut. One thing that could lead to tire damage on takeoff or landing is a bad or improperly serviced strut.

The caption on the back of this image reads, "A U.S. Navy North American T-2C Buckeye (BuNo 156712) from NAS Meridian, MS, where jet basic training was conducted" with a North American T-39D Sabreliner (BuNo 150980), both of Training Squadron (VT) 10. The planes are in flight near Naval Air Station Pensacola, Florida. A former instructor at NAS Pensacola stated, "Every pilot went through primary training at Saufley Field then entered either the propeller or jet pipeline after completing primary training."

Four

THE BLUE ANGELS

The Blue Angels, the Navy's premier flight demonstration team, are known the world over for their gleaming aircraft and breathtaking aerobatics. The Blue Angels call Pensacola home from March to November, and their shows and practices in Pensacola are attended by hundreds of thousands every year. In 1946, Adm. Chester Nimitz, at that time chief of naval operations, wanted a team to showcase the power of naval aviation and the skill of her pilots and crew members. The team was originally led by Capt. Roy Marlin "Butch" Voris, a World War II flying ace whose abilities in the cockpit earned him three Distinguished Flying Crosses and the reputation of being "a fighter pilot's fighter pilot." When asked about how close they fly, he would respond, "if we hit each other, we're too close, and if we don't, we're too far apart."

The team's first show was given on June 15, 1946, in Jacksonville, Florida. It lasted 17 minutes and featured nine maneuvers, including the Cuban Eight, an inverted pass, and the "Zero" routine, developed by Voris to simulate an aerial dogfight.

Originally based in Jacksonville, and then Corpus Christi, the Blues have called Pensacola home since 1954. From March to November, visitors can see them practice at Sherman Field. For times and days of practice, see www.blueangels.navy.mil. From November to March, they spend their time at El Centro, California.

Today, 11 million fans flock to see the Blues perform each year. While their official mission, as stated on their website, is to "enhance Navy recruiting, and credibly represent Navy and Marine Corps aviation to the United States and its Armed Forces to America and other countries as international ambassadors of good will," they also leave their fans thrilled and amazed, whether witnessed for the first time or after many times.

This is a photograph of the original crew of the Navy's Flight Exhibition Team in 1946, posing in front of their Grumman F6F Hellcats. Shown here are, from left to right, Lt. Al Taddeo, Lt. (jg) Gale Stouse, Lt. Comdr. "Butch" Voris, Lt. Maurice Wickendoll, and Lt. Mel Cassidy.

Team leader Butch Voris sits in the second Blue Angels aircraft, the Grumman F8F Bearcat, in Jacksonville in 1946. During World War II, Voris flew from the carriers *Enterprise* and *Hornet* and participated in such famed battles as Tarawa, Coral Sea, and Philippine Sea. He passed away in 2005, but his legacy and influence remain today.

Voris poses with his Bearcat in 1946. The Bearcat was "the last propeller-driven Navy fighter," according to the NNAM website. There is a fully restored Bearcat on display at the museum.

After its first performance, the team was awarded this trophy for "the Finest Exhibition of Precision Flying" at the Southeastern Air Show and Exhibition. The trophy is now kept in a place of honor in the ready room of the Blue Angels headquarters at NAS Pensacola.

The 1947 team poses in front of their Bearcat and a 1946 Buick Super. Shown here are, from left to right, "Butch" Voris, "Wick" Wickendoll, Bob Clarke, Al Taddeo, and Billy May.

The Blue Angels watch the diamond formation performed by pilots in F9F-2 Panthers, the Blues' first jet-powered aircraft, in 1949. It is to the credit of the crew members that the Blue Angels have never had to cancel a performance due to mechanical failure.

The caption on this photograph states, "A practice flight in the new Grumman F11F-1 Tigers over the Pensacola area. The large tail numbers were not added until after the first few air shows of the 1957 season." The Tiger was the Blues' first supersonic jet.

"Pilots are seen waving from the new [Grumman F9F-8] Cougar planes," in this 1956 photograph. One change in the Cougar from the F9F Panther was a swept wing instead of a straight wing.

From 1969 to 1974, the Blues flew the McDonnell Douglas F4J Phantom. The Air Force Thunderbirds flight demonstration team flew the same aircraft, the only plane to hold that dual distinction.

The Douglas A-4F Skyhawk was flown by the Blues from 1974 to 1986. Its compact size allowed it to continue to fly from the smaller World War II–era carriers.

Blue Angels pilots pose in the early 1970s. They are, from left to right, Steve Lambert, Bill Switzer, Mike Murphy, Don Bently, Bill Beardsley, Skip Umstead, and Gary Smith.

Bob Hope (center) poses with the crew of the 1986 plane A-4 Skyhawk. Hope was made an honorary member of the Blue Angels in 1982, including a brief flight in the Skyhawk. After the flight, he said, "It was the smoothest ride I have ever taken," and "It's like owning some shares of heaven when you're up there." Read more about his experience at www.blueangels.org.

This photograph of the team features the first African American Blue Angel, Lt. Comdr. Donnie Cochran, posing in front of their current jet, the McDonnell Douglas F/A-18 Hornet. Shown here are, from left to right, Lt. Mike Campbell, Lt. Wayne Molnar, Lt. Comdr. Pat Walsh, Capt. Mark Birchner (USMC), Lt. Comdr. Gil Rud, Cochran, Lt. Comdr. David Anderson, and Lt. Cliff Skelton.

Throughout the years, the Blue Angels have employed a number of support aircraft, including the Douglas R4D Skytrain, the Curtiss R5C Commando, the Douglas R5D Skymaster, and the Lockheed C-121 Super Constellation. Today, a Marine Corps crew flies the Lockheed C-130 Hercules (pictured), fondly known as "Fat Albert." The Hercules, with its 132-foot wingspan and 45,000-pound payload, supplies the crew of the Blues with everything necessary to keep the planes in the air. The crowds are never disappointed. (Courtesy of Public Affairs Office, Blue Angels.)

Five

NAS PENSACOLA THEN AND NOW

The first lasting military establishment in the area was built on the bluffs overlooking Pensacola Bay in 1698 by the Spanish. The French battled the Spanish, the Spanish battled the British, and the British battled the Americans. The British enlarged on the facility, and the US Navy began building on that spot in 1826. In 1837, the Navy began constructing a wall to enclose the yard. It is believed that slave labor was used in part to build it, as George F. Pearce, in his book *The U.S. Navy in Pensacola*, reports on an advertisement for laborers in the *Pensacola Gazette* of May 13, 1837: "While the advertisement did not specify slave laborers, it implied as much: 'The owners . . . [will] subsist and quarter them beyond the limits of the Yard.'"

The largest military action took place during the Civil War, when Confederate and Union forces engaged in a shelling contest across the bay. The Confederate forces abandoned the base after the fall of New Orleans, but destroyed most of the buildings in their retreat. The National Register of Historic Places conducted a survey of the base in 1976, giving a good deal of historical and architectural information on the older buildings, but many of those buildings have since been lost. Since that time, the area has been hit by dozens of hurricanes causing considerable destruction, including Hurricane Ivan in 2004. As a result, the search for old buildings on the base can be disappointing. Here are some old and new photographs of the buildings that are still standing, with a bit of their history.

This 1915 aerial photograph shows the new Naval Aeronautical Station, before Station Field (later Chevalier Field) was built. The old towns of Woolsey on the east and Warrington on the west are still visible. The residents of Warrington were relocated, and Woolsey was leveled as the Navy expanded its aeronautic activities.

Almost a century of building can be seen in this more recent aerial photograph of NAS Pensacola, taken during a visit from the USS *John F. Kennedy* (CVN-67). (Courtesy of Public Affairs Office, NAS Pensacola.)

The Spanish built fortifications and other buildings, including a chapel, beginning in 1698. All the buildings were made of wood, but footings and material remains have been excavated and examined by the Archaeological Department of the University of West Florida beginning in 1995. For more in-depth reading, see *Presidio Santa Maria de Galve: A Struggle for Survival in Colonial Spanish Pensacola* by Dr. Judith Bense. (Courtesy of University of West Florida Archaeology Institute and NAS Pensacola.)

A partial reconstruction of the palisade of Fort San Carlos de Austria is located on Slemmer Road just across from NAS Pensacola headquarters. A great exhibit of artifacts excavated from the dig can be found at the Archaeological Institute on the campus of the University of West Florida. (Photograph by the authors.)

Bateria de San Antonio is part of the Fort Barrancas complex. It was finished in 1797 and is one of the oldest extant fortifications in Florida. The Spanish built it after seizing the Royal Navy redoubt from the British in 1781. (Courtesy of Gulf Islands National Seashore.)

In 1839, the Americans built Fort Barrancas on the same site, preserving for future generations this reminder of Pensacola's heritage of "Five Flags." Both Fort Barrancas and Fort Pickens are excellent examples of fortification architecture. (Photograph by the authors.)

Building 191 was not originally part of the Navy base or the air station, but was known as the Warrington Store. According to the Commander, Navy Installations Command (CNIC) Office, it was the "grocery store and home of Fred Bauer, constructed in the early 1850s." Indeed, the 1880 federal census for Warrington lists Fred Bauer, a Bavarian immigrant, as a retail grocer with his family. (Courtesy of Personnel Support Activity Detachment Building.)

After being used for multiple purposes over the years, the old Warrington Store is now Building 191. It houses the Navy-Marine Corps Relief Society, located on Radford Boulevard. (Photograph by the authors.)

In 1837, the Navy began building a 10-foot-high wall around the approximately 80 acres that had been dedicated for the yard, complete with gates. In his book *The U.S. Navy in Pensacola*, George F. Pearce relates that because stone was difficult to acquire in Florida, the Navy commissioner allowed Commandant Bolton to use brick on the wall, provided it was "very hard burnt brick." Advertisements for laborers were placed in newspapers as far away as New Orleans, Mobile, Natchez, and Montgomery.

This view is looking east at Radford Boulevard and West Avenue. A civilian Navy yard worker in the mid-to-late 1800s to early 1900s who left the yard by this gate would be headed home to the village of Warrington. Today, he or she would be driving down Radford Boulevard, passing the Naval Aviation Schools campus, where water survival, aircrew, and aviation safety are taught. Farther on are the Mustin Beach Pool and Fort Barrancas. (Photograph by the authors.)

Workers and their family members from the Navy yard who lived in Warrington and Woolsey were buried in a cemetery just north of the old naval hospital. A glimpse of the 1850 federal census of Warrington shows a preponderance of skilled workers, such as ships' carpenters, joiners, blacksmiths, stonecutters and masons from New England, Canada, and Ireland, with only a few native-born Floridians.

As the aeronautic station grew and Woolsey and Warrington were absorbed, the cemetery was relocated in the 1930s, with most of those interred in the Old Warrington Cemetery moved near Barrancas National Cemetery. Today, the Old Warrington Cemetery has been absorbed by the national cemetery. The old cemetery can be found at a group of old headstones surrounded by small shrubs. (Photograph by the authors.)

The Pensacola Lighthouse is one of the oldest lighthouses in continuous use in Florida today. It was built in 1858 and, despite wars and hurricanes, still shines its light over Pensacola Bay, Santa Rosa Island, and Perdido Key.

The lighthouse is open for daily and special tours, where visitors can enjoy a spectacular view after climbing 177 steps to the top. Many special tours are offered at the lighthouse, such as ghost tours and Blue Angels watching events. The view looks out over Pensacola Bay. (Photograph by the authors.)

Building 16 has been repurposed multiple times since its construction in 1834, serving as an armory, a chapel, and an officers' club. Because of its use as a chapel in 1862, it was spared destruction by the retreating Confederates. It is on the list of "haunted" places in Pensacola, as many who have worked there have reported hearing poker chips being dropped, reportedly by the ghost of a Marine Corps pilot who spent his downtime playing poker at the officers' club.

Building 16 recently underwent a major renovation, including the wraparound porch, and now houses trial services for the legal department. (Photograph by the authors.)

Completed in 1835, the naval hospital was surrounded by a 14-foot wall. Legend has it that locals believed the height of the wall would keep out mosquitoes, but, as the historical sign suggests, it was not until 1893 that the mosquito was identified with malaria.

All that remains of the hospital are a section of the wall and gate, and the steps up the hill. The area where the hospital once stood is now enlisted housing. The current Naval Hospital Pensacola was built off-base in 1976. (Photograph by the authors.)

In 1852, construction was begun on a granite-block wet basin. This 1918 photograph shows new life rising from the hurricane-battered yard with the construction of a naval air station. Note the camouflage paint scheme on Building 38.

Today, the wet basin echoes with the sound of modern motors. Even Air Force ships can be found there. According to the Historic American Buildings Survey (HABS), the basin is used today for "the berthing and repairing of small craft." (Photograph by the authors.)

Built in 1868, the ship's carpenter's workshop has since become another victim of a hurricane's wrath.

Today, only the smokestack remains, standing guard over a park built at its base. (Photograph by the authors.)

Quarters A, the commandant's quarters, was built in 1874 on the same site as the previous quarters, which had been erected in 1826 and burned by the Confederates in 1862. The caption reads, "A well-dressed Naval officer in front of quarters, ca. 1880–1885." This officer is clearly out of uniform. The base commandants for that time period would have been George E. Belknap, S.P. Quackenbush, William Welch, Robert F. Bradford, William C. Gibson, E.P. Lull, and W.M.C. Gibson

Today, Quarters A is home to the commander of the Naval Education Training Command. In the past, notable officers, such as Adm. William Halsey and Adm. Chester Nimitz have called Quarters A home. (Photograph by the authors.)

Behind Quarters A and parallel to the original wall is a pair of buildings once used as stables for the officers' horses, in the days when the US Army maintained Fort Barrancas as an active-duty post. At the time, the cavalry still rode horses.

The stables, now Buildings 19 and 28, were built in 1874. On the other side of the wall from the stables was Chevalier Field and its hangars. Today, Saufley Street runs along the wall, with the Navy Exchange, Portside Cinemas, and Portside Fitness in the old hangars.

"With the establishment of an aeronautics station, for training purposes, lighter-than-air was one of the branches of instruction. This is a picture of a free balloon ready for ascent, taken in front of building 18, April 15, 1916," according to the photographer's caption.

Building 18 is long gone, but its twin, Building 38, "sits on the site of the 1829 blacksmith shop," according to the National Register of Historic Places in 1972. Building 38 was equipped with schools for apprentices and elevators for lifting aircraft engines to the second floor. (Photograph by the authors.)

Naval and Marine Corps officers pose in front of the "Flying School Office" around 1917. A few have been identified by the researchers at the Buehler Library. The three men standing on the left are, from left to right, H.W. Scofield, unidentified, and C.K. Bronson. The two seated on the second step are unidentified (left) and Earl W. Spencer. The three standing on the right are, from left to right, W.M. Corry, Bernard L. Smith, and Wadleigh Capehart. The four men seated are, from left to right, M.L. Stolz, Robert Paunrack, Earl F. Johnson, and unidentified.

Things have come a long way. This is the home of the Naval Education Training Command, whose mission is to "be the global leader in rapid development and delivery of effective, leading edge training for naval forces." (Photograph by the authors.)

"Planes inspection and review, Chevalier Field" are shown in this photograph. The planes appear to be the Curtiss F8C-4. If so, this photograph would have been taken sometime from approximately 1925 to 1935. Note the officers' housing on Billingsley Road in the background.

Today, the officers' housing is recognized by the National Register of Historic Places as "a fine example of regionally adapted architecture." Under swaying Spanish moss, they stand on a hill overlooking what is now Chevalier Hall and Pensacola Bay. They were built from 1874 to 1876. (Photograph by the authors.)

This 1919 photograph shows the foundry (left) and the electric shop, with its camouflage paint scheme. By the time of the Historic American Buildings Survey, done in 1972, the electric shop was gone.

Today, only the foundry remains. It was built in 1882. If its walls could speak, they would tell a story of the constant change of military technology, of going from casting parts for sailing and steamships to parts for wooden and, later, aluminum aircraft. Many of the 19th-century buildings still standing are now employed for storage and other mundane uses, mere shadows of their former function. (Photograph by the authors).

The caption accompanying this August 1949 photograph reads, "An SNJ-5 assigned to the Blue Angels Flight Demonstration Team pictured at Chevalier Field on board NAS Pensacola." Behind the SNJ-5 is one of the brick hangars built in the 1920s.

Although the hangar now houses a basketball court, the elevation marking "Elev. 9 Ft." still proclaims the proximity of NAS Pensacola to sea level. (Photograph by the authors.)

The caption on the back of this image reads, "With the filling of the Commodore's Pond in 1929, the modern air station lost its last link with the Navy of Wooden Ships. Live oak for the construction of ships was stored in this pond, the earliest found dated 1812. With the advent of steel ships, it remained for years as it is seen here, some logs above water, thousands of board feet submerged in this storage. When the Frigate Constitution was reconstructed, most of the wood used in her repair was live oak logs taken from this pond. Today, this filled area lies for the most part under Chevalier Field."

What was then the location of Chevalier Field is now the site of Chevalier Hall, housing the Naval Aviation Technical Training Command (NATTC). (Photograph by the authors.)

In this reproduction photograph, reprinted in 1964 for the 50th anniversary of NAS Pensacola, a Curtiss H-12L sits in the hangar area. The original photograph was taken in 1918.

The seaplane hangars are quiet now. Except for pleasure boats, most of the noise today occurs at Forrest Sherman Field. (Photograph by the authors.)

An inspection is carried out near the hangars in 1918, likely part of demobilization of what had grown to be one of the largest of the world's navies during World War I. Adm. William Benson, the Navy's chief of naval operations during the war, oversaw the growth of the Navy, but wanted to rid it of aviation, as he did not "conceive of any use the fleet will ever have for aviation." Fortunately, the assistant secretary of the Navy, Franklin D. Roosevelt, had other ideas.

Originally, as many as 10 of the metal hangars were built, but now, all but two have been lost to time and hurricanes. In the 1970s, the Navy had plans to demolish the remaining hangars and build a greenbelt along the water, but a 1972 survey by the National Register of Historic Places seems to have saved Buildings 73 and 74. The survey described them as "the most significant buildings from the standpoint of aviation history." (Photograph by the authors.)

In 1918, this area was one of the busiest on base, with a dozen different aircraft in various stages of launch, recovery, maintenance, and repair. The two seaplanes in the water near the center of the photograph are Curtiss R-6s, and the larger aircraft to the right on the deck is a Curtiss H-12.

Today, only the tracks of the huge hangars remain to remind visitors of that era. The asphalt provides parking space for nearby buildings. Part of the beach area that still includes the old cement seaplane ramps now features the Desert Storm Memorial. (Photograph by the authors.)

An early gate welcomes the visitor to NAS Pensacola, the naval hospital, and Fort Barrancas. The fort remained an active US Army post until advancements in artillery rendered brick fortifications completely ineffectual. The Army surrendered the fort to the Navy in 1946.

Today, Duncan Road is paved, and a larger, more modern facility stands guard to the main entrance to NAS Pensacola. (Photograph by the authors.)

106

Building 633, Naval Ground School, is seen here in 1942. In his excellent paper "Striking Eagles: Doctrine, Training, and Fighting Squadron Five at War in the Pacific," Michael Elliot Kern points out that "in early 1942, the Navy fighters were the most maligned branch of naval aviation and were completely outclassed by their Japanese opponents. By 1944, the fighters had become the most numerous and important type of aircraft on the aircraft carrier's flight deck," a testament to the training that occurred here.

Building 633 looks very much the same today, except for the vehicle parked in front. Now, the building houses the campus of Naval Aviation Schools Command. (Photograph by the authors.)

Built in 1939 as an Army barracks, this beautiful building overlooking Pensacola Bay later housed the Navy's Photography School. This photograph shows a class from 1968, complete with female sailors—no longer Yeomanettes or WAVES, but sailors in their own right.

Located about 1,500 feet east of Fort Barrancas, this building was an Army barracks in 1939. Today, Building 1500 houses the headquarters of the naval air station. A small museum of naval photography is also housed there, but is not open to the general public. (Photograph by the authors.)

Building 34 is described by the National Register of Historic Places as "surrounded by a two-tiered veranda supported by iron columns, [the] white-trimmed, two-story, tan brick quarters" was originally built in 1873 as the station's bachelor officers' quarters. The Historic America Buildings Survey (HABS) states it matches the Marine Hospital in Galena, Illinois, which was designed by early American architect Ammi B. Young.

Today, Building 34 is used as a guest house for Navy Gateway Inns and Suites, and as such is only open to military members and their families. But the sight of it, surrounded by live oaks draped with Spanish moss, is a reminder of the rich naval and Floridian heritage that still flourishes here.

Newly commissioned officers typically pose in front of the Naval Aviation Memorial Chapel after the ceremony. This group is Aviation Officer Candidate School Class 04-81, in front of the station chapel on May 1, 1981. At far right, second row is class officer Lieutenant Allbee, USN. Ens. Randall W. VerHoef (fourth row, fifth from left) became an intelligence officer, serving with HS-8 on the USS *Constellation* and the USS *Ranger*; with Fleet Ocean Surveillance Information Facility, Rota, Spain; and with VQ-2. At left is Gy. Sgt. Buck Welcher (USMC). Shortly after this photograph was taken, Welcher served as technical advisor on the 1982 film *An Officer and a Gentleman*, Hollywood's version of the AOCS experience. (Courtesy of VerHoef family.)

The Naval Aviation Memorial Chapel was built in 1961, a half-century after the establishment of naval aviation, according to the CNIC Office. The chapel celebrated its jubilee anniversary in May 2011 with concerts and renewal of marriage vows for couples who had been married there. (Photograph by the authors.)

Six

NAS PENSACOLA TODAY

In spite of hurricanes, periods of inactivity, and even rumors of BRAC (Base Realignment And Closure) lists, more than 100 years after its founding as a naval air station, NAS Pensacola is a bustling hub of aviation and education. According to the base website, the base "provides support to 89 Department of Defense (DoD) and 30 non-DoD tenant commands, most of which are primarily dedicated to the training of Navy, Marine Corps, and Coast Guard personnel in naval aviation," and "employs more than 16,000 military and 7,400 civilian personnel. This includes major tenant commands Naval Air Schools Command, Naval Air Technical Training, Marine Aviation Training Support Group 21, the Blue Angels Navy flight demonstration squadron, and the headquarters site for CNET, a command which combines direction and control of all Navy education and training."

In addition to being an active military base, NAS Pensacola is also a major tourist destination. The National Naval Aviation Museum is the largest of its kind in the world. In the last 10 years, its "visitation has more than doubled, approaching one million in some years," as stated on its website. Barrancas National Cemetery is an active military cemetery, with over 30,000 burials. Fort Barrancas and its Advanced Redoubt are excellent examples of 19th-century fortifications. Pensacola Lighthouse is one of Florida's oldest operating lights. The base's biggest draw is the Blue Angels, whose practices at Forrest Sherman Field are open to the public from March to November.

Naval aviator Capt. Christopher Plummer took command of NAS Pensacola in April 2010. In March 2013, he turned command of the station over to former Blue Angels officer Capt. Keith Hoskins. (Photograph by the authors.)

A Blue Angels F11F-1 Grumman Tiger from the late 1950s welcomes the visitor just inside the east gate. Other F11F models are on display in museums in Canton, Ohio; Topeka, Kansas; Tucson, Arizona; Garden City, New York; Lakehurst, New Jersey; Oceana, Virginia; and Kalamazoo, Michigan. (Photograph by the authors.)

Seen here is Training Air Wing Six (TRAWING 6), whose commander informed the authors that the wing "trains and graduates approximately three hundred United States Navy, Marine Corps, and international students annually. Students from Germany, Italy, Saudi Arabia, and Singapore represent approximately ten percent of the total." Aircraft such as T-39 Sabreliners, T-6A Texan IIs, and T-45 Goshawks are flown by trainees in the course of their instruction.

Chevalier Hall is home to the Naval Aviation Technical Training Command (NATTC). The mission of NATTC, as stated on its website, is to "produce the most knowledgeable, best-trained Sailor today." The command, established in 1942, "has grown from three schools to the present 110 courses" and "graduates approximately 15,000 Navy and Marine students yearly."

Designated a national cemetery in 1868, Barrancas National Cemetery possesses an aura of history and peaceful repose. The National Park Service (NPS) reports that "the double-iron gate at the main entrance, at the southern edge of the cemetery, dates to 1868, and the nearby pedestrian gate dates to 1936. A portion of the original brick wall enclosing the cemetery remains along the site's western boundary, though wrought-iron fencing replaced the wall in other areas." There are several Medal of Honor recipients interred there, as well as Geronimo's second wife, Ga-Ah, and many "Colored Infantry" troops from the Civil War.

The Advanced Redoubt is also administered by the NPS. It was built near Fort Barrancas to defend the Navy yard from land assault. Because of that purpose, according to the NPS, it is "unique among the early American forts at Pensacola in being designed solely for resisting a land-based assault."

Today, Fort Barrancas is cared for by the National Park Service. Visitors can cool off in the brick tunnels, feel the echoes of hundreds of years of history, and study fort construction from Spanish, British, and American eras. (Photograph by Kaiden S. Keillor.)

Pensacola Lighthouse is one of Florida's oldest lighthouses, and it continues to be operated by the US Coast Guard. Its 177 steps can be climbed by visitors to enjoy a spectacular view of the sparkling white beaches, Santa Rosa Island across the bay, and the air station. Special tours, such as the Light of the Moon tour, Blue Angels Practice, Toast of the Top Sunset tour, and ghost hunts are available. It is, after all, one of Pensacola's most haunted places!

The National Naval Aviation Museum can be seen to the northeast from the top of the lighthouse. The 50th anniversary of the museum is being celebrated in 2013. The museum began in a little wooden hut left over from World War II. Today, it is "one of the most-visited museums in Florida," with attendance in some years approaching one million, as stated on its website.

Across the Pensacola Bay to the south is Santa Rosa Island and Fort Pickens. Fort Pickens on Santa Rosa, Fort McRee on Perdido Key, and Fort Barrancas on the mainland formed a deadly triangle to any invading ships and protected Pensacola's valuable deepwater harbor.

Sherman Field and Training Wing Six headquarters can be seen in this north-facing view. Frequently, visitors to the lighthouse are treated to flyovers from jets taking off, landing, and performing touch-and-goes, as the business of pilot training continues.

Perdido Key, where Fort McRee once stood, is seen in this southwest view. Union forces retreated from Fort McRee, and it was occupied by forces from Alabama and Florida. It was then bombarded by Union forces holding Fort Pickens, was badly damaged, and was later burned by retreating Confederate forces. Little was done to repair or maintain the fort for many years. During both world wars, the Army began to build batteries at the site, but McRee never saw action again. Erosion and storms have washed away what remained of the original brick fortification, but 1943's Battery 223 and its concrete footings can still be seen. However, it is only accessible by a long walk along the beach or by boat.

Forrest Sherman Field, the home of the Blue Angels, is also home to Training Air Wing Six, where thousands of pilots from the Navy and Marine Corps earn their wings every year. According to the TRAWING 6 website, students from Germany, Italy, Saudi Arabia, and Singapore also train there.

National Naval Aviation Museum was originally opened in a World War II–era wood building in 1963. Today, it houses hundreds of aircraft in 300,000 square feet, plus several aircraft outside on the grounds. Its mission, as stated on its website, is "To foster and perpetuate the Naval Aviation Museum as a medium of informing and educating the public on the important role of United States Naval Aviation and to inspire students undergoing naval flight training to complete the program and become career officers; to inspire young people to develop an interest in Naval Aviation."

These Douglas A-4 Skyhawks, once flown by the Blue Angels, hover 75 feet above the floor in this beautiful glass atrium. They are only a few of the approximately 150 aircraft on display, representing every era of naval aviation.

Every aircraft in the museum is painstakingly repaired and polished until its shines like new. As stated on the museum's excellent website, "The mission of the National Naval Aviation Museum, an official Department of the Navy museum, is to 'select, collect, preserve and display historic artifacts relating to the history of Naval Aviation.'"

This Stearman N2S-3 Kaydet was flown by Pres. George H.W. Bush during his flight training as a cadet. Also on display is his logbook. Bush's 58 missions as a naval aviator during World War II earned him three Air Medals and the Distinguished Flying Cross.

One of the very first carrier jets, the F9F-2 Grumman Panther, proved itself in the Korean War. It was flown by the US Navy from 1947 to 1958. The single-seater was powered by one Pratt & Whitney J48-P-8 engine.

This former ground-support equipment from Sherman Field has been repurposed to help visitors reach the second floor of the museum. The second floor contains an excellent balloon exhibit as well as several interactive cockpits for children to experience.

This Curtiss NC-4 was the first aircraft to successfully cross the Atlantic, completing the voyage in 1919. It is on loan from the Smithsonian Institution. This behemoth featured a wingspan of 126 feet and was powered by four Liberty V-12 engines. Its 1,600-horsepower engines required 1,291 gallons of gasoline to generate the power to carry its crew across the Atlantic.

The main floor of the museum includes the island and flight deck of the USS *Cabot* (CVL-28), the last survivor of the light aircraft carriers of World War II. Among the many historically significant aircraft on display are the N2S-3 Kaydet flown by Aviation Cadet George H.W. Bush in 1943 and a SBD-2 Dauntless that survived Pearl Harbor and Midway.

A nicely restored 1940 Plymouth welcomes visitors at the back door of the museum. It is a great segue to a second-floor exhibit, *Home Front*, detailing home life during World War II. There is also an exhibit about life in the Pacific theater during that time.

Hangar Bay One, opened in 2010, adds 55,000 square feet of floor space to exhibit such aircraft as Marine One, from the Nixon-Ford era; P5M Marlin, the Navy's last seaplane; a Cessna O-1 Bird Dog, flown out of Saigon in 1975 by a South Vietnamese Air Force pilot carrying his wife and five children, and landed on the USS *Midway* (CVA-41); and the SBD-3 Navy One, which landed on the USS *Abraham Lincoln* (CVN-72), carrying Pres. George W. Bush. There is also a special prisoner-of-war exhibit in Hangar Bay One.

This view of one corner of Hangar Bay One shows an LTV A-7 Corsair cockpit (left), an F-8 Crusader (center), and a McDonnell Douglas F/A-18 Hornet (right). Hanging from the ceiling is the "Truculent Turtle," a P2V-1. In 1946, the Turtle broke the record for unrefueled flight, going from Perth, Australia, to Columbus, Ohio, a distance of 11,236 miles, in 55 hours and 18 minutes. (Photograph by Kaiden S. Keillor.)

A retired Blue Angels F/A-18 Hornet has a new home in Hangar Bay One. Every visit to Hangar Bay One brings a new experience. A recent exhibit, *From Typewriters to Strike Fighters: Women in Naval Aviation*, features video, interviews, and artifacts detailing the career of women in the Navy—from Yeomanettes to WAVES to fighter pilots.

The National Flight Academy, a subsidiary of the National Naval Aviation Museum, was authorized by the Department of the Navy in 2001. Its mission is to "inspire students," particularly in the areas of science and technology. "Disciplines include aerodynamics, propulsion, navigation, communications, flight physiology and meteorology, along with core values, teamwork, and leadership skills development." The University of West Florida assisted with curriculum development, and seventh-through-twelfth-grade students can live in quarters while they learn and can attend summer camps.

A short walk from the lighthouse takes the visitor to a boardwalk leading to a lovely beach area. With Santa Rosa Island in view across the bay, it is a perfect spot for Blue Angel spotting.

On special occasions, vintage aircraft can be seen doing flybys, such as this B-29 Superfortress, the largest aircraft used in World War II. About two dozen B-29s are known to be in existence. "Fifi," of the Commemorative Air Force, based in Texas, is the only known craft to be flightworthy. (Courtesy of Scott Slocum of AeroMedia Group.)

The "Mighty O," USS *Oriskany* (CVA-34), was a World War II–era carrier launched in New York in May 1944. After earning two battle stars in Korea and five in Vietnam, she was decommissioned in 1975. The decision was ultimately made to scuttle her off the coast of Pensacola as a living reef and as a memorial to the men who sailed with her. She is the largest ship ever sunk for that purpose.

In 2004, the Mighty O arrived at NAS Pensacola. In May 2006, she was towed 24 miles out and sunk with strategically placed explosives. She sank in 37 minutes. She has become the world's largest artificial reef, and was named in 2007 by the *Times* (London) as one of the "top ten wreck-diving sites in the world." The *Oriskany* lies in 200 feet of water, so diving throughout the ship is for more experienced divers. A May 2011 article in *Scuba Diving* magazine states that "the ship's superstructure rises to 80 feet and is a divers' playground." (Courtesy of Oriskany Museum of Oriskany, New York.)

Today, the trip from NAS Pensacola to Fort Pickens is a 25-mile drive. In the near future, a ferry will operate from Pensacola to Santa Rosa Island, where visitors can learn more about fortification-building in various stages, from the early 19th century to the mid-20th century. This photograph from Fort Pickens shows one of the few remaining Rodman 15-inch cannons standing guard over an American 19th-century fortification.

Driving out of the west gate from NAS Pensacola, Perdido Key can be reached in a few minutes. It is home to Rosamond Johnson Jr. Beach, another component of the national park system. During the Jim Crow era, Pensacola Beach was reserved for whites only, and Perdido Key was open to African Americans. In 1950, Army private Johnson was killed at age 17 in the Korean War while saving two fellow soldiers. The beach was later named in his honor. A move is ongoing to see that he is properly honored by the military for his sacrifice.

DISCOVER THOUSANDS OF LOCAL HISTORY BOOKS FEATURING MILLIONS OF VINTAGE IMAGES

Arcadia Publishing, the leading local history publisher in the United States, is committed to making history accessible and meaningful through publishing books that celebrate and preserve the heritage of America's people and places.

Find more books like this at
www.arcadiapublishing.com

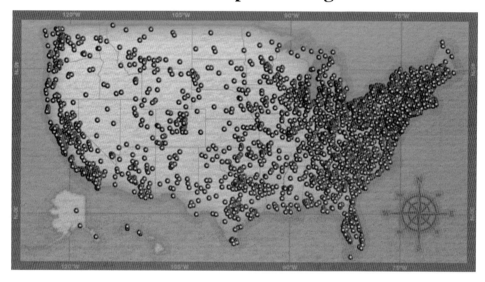

Search for your hometown history, your old stomping grounds, and even your favorite sports team.